MW00888262

FROM **CLAY** TO **BRICKS**

by Stacy Taus-Bolstad
photographs by Cheryl Richter

Lerner Publications Company / Minneapolis

Lerner Publications Company
A division of Lerner Publishing Group
241 First Avenue North
Minneapolis, MN 55401 U.S.A.

Website address: www.lernerbooks.com

Library of Congress Cataloging-in-Publication Data

Taus-Bolstad, Stacy.
From clay to bricks / by Stacy Taus-Bolstad.
p. cm. — (Start to finish)
Includes index.
ISBN: 0-8225-4663-9 (lib. bdg. : alk. paper)
1. Brickmaking—Juvenile literature. [1. Brickmaking. 2. Bricks.] I.Title. II. Start to finish (Minneapolis, Minn.)
TP827.5 .T38 2003
666'.737—dc21 2002006597

Manufactured in the United States of America
1 2 3 4 5 6 - JR - 08 07 06 05 04 03

Table of Contents

Bricks make buildings strong.

How are bricks made?

Workers find clay.

Bricks begin as clay. Clay is a thick, heavy kind of dirt. Workers find clay by digging a large pit called a **quarry**.

5

Workers dig up the clay.

Workers use huge machines to dig clay from the quarry. The workers put the clay in trucks. Trucks take the clay to a **factory** to be made into bricks.

Machines crush the clay.

Workers at the factory let the clay sit until it is dry. Then machines crush the clay into very tiny pieces. Workers add water to the crushed clay to make it as thick as stiff mud.

The clay is shaped.

The clay is squeezed through a **mold**. A mold is a part of a machine. The mold shapes the clay into a long, thick ribbon.

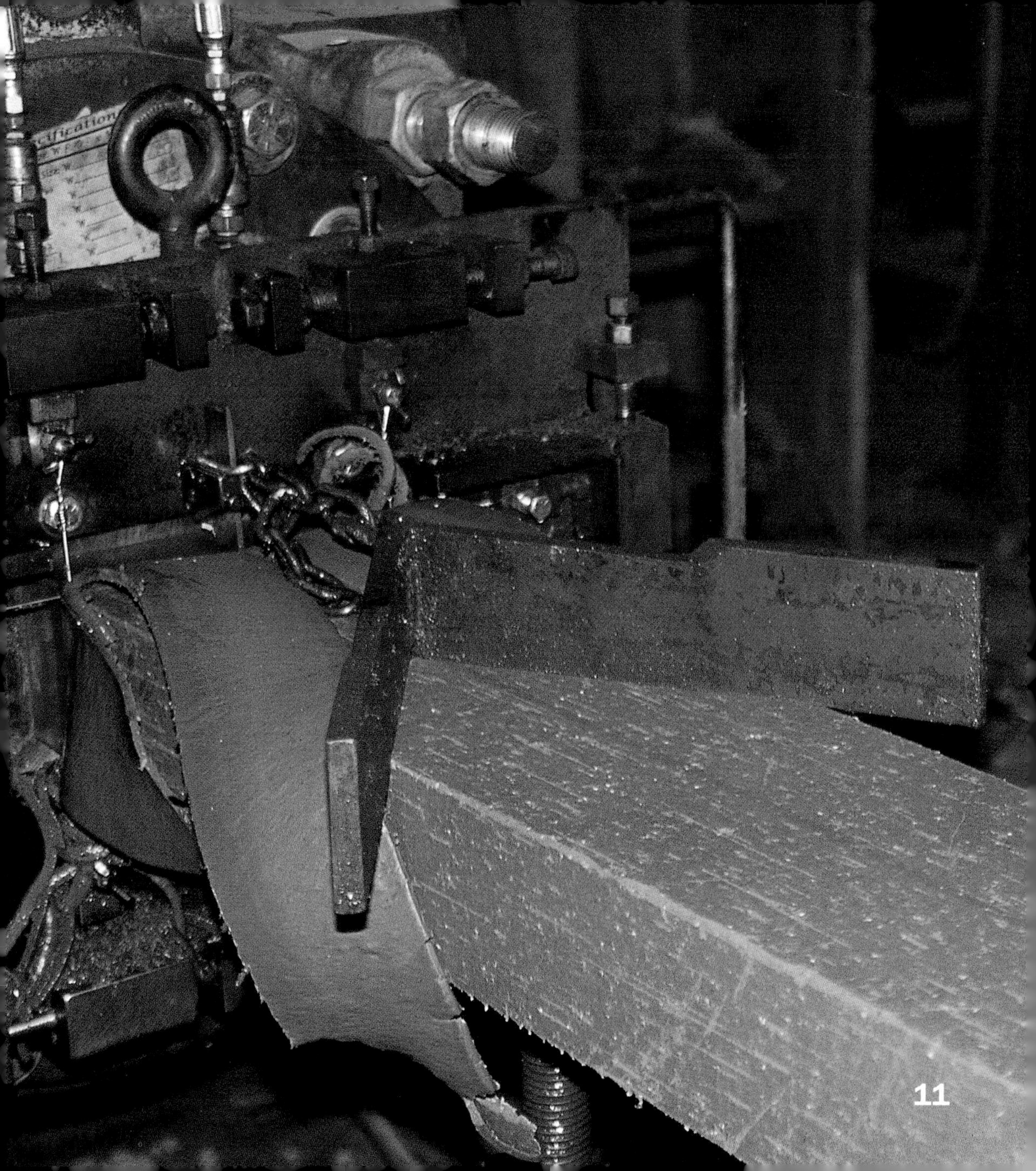

Wires cut the clay.

Wires cut the clay ribbon into bricks. The bricks are stacked in big piles.

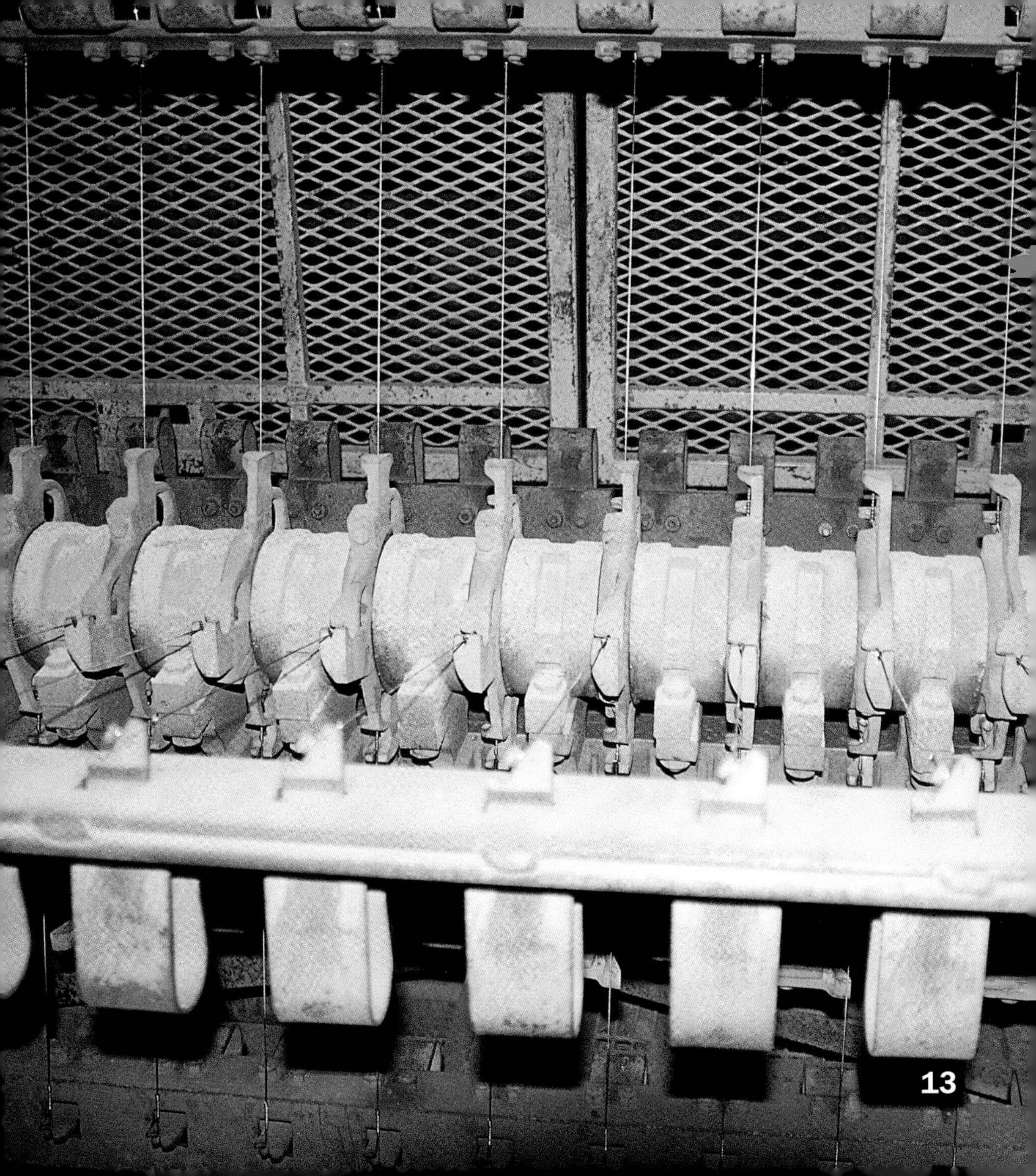

Warm ovens dry the bricks.

The bricks are still wet. They must be dried to make them hard and strong. Workers set the bricks on carts. The carts are put into warm ovens to dry the bricks.

Hot ovens bake the bricks.

Next the carts pass through a big **kiln**. A kiln is a very hot oven. The kiln bakes the bricks to make them even stronger.

Workers sort the bricks.

The bricks leave the kiln and cool down. Workers sort the bricks. Bricks that are broken or twisted are thrown away.

Workers pack the bricks.

Workers put the bricks in stacks. The stacks are loaded onto trucks. The trucks take the bricks to builders.

HYSTER
HYSTER 65

Builders build with bricks.

Builders use the bricks to make homes and schools. Is your home made of bricks?

Tyvek

Glossary

factory (FAK-tuh-ree): a place where clay is made into bricks

kiln (KIHLN): a very hot oven used to bake bricks

mold (MOHLD): a part of a machine that shapes clay into a long ribbon

quarry (KWOHR-ee): a pit where workers dig clay

Index

DATE			